To our readers, those students just beginning a life of adventure and learning, always remember:

*You are braver than you believe, stronger than you seem,*
*and smarter than you think.*
*~*
A.A. Milne

Written by


# Brent A. Ford

# &

# Lucy McCullough Hazlehurst

I wonder what I'll look like when I grow up.

You'll grow up to
be that gentle giant
of the jungle,
an elephant.

I wonder what I'll look like when
I grow up.

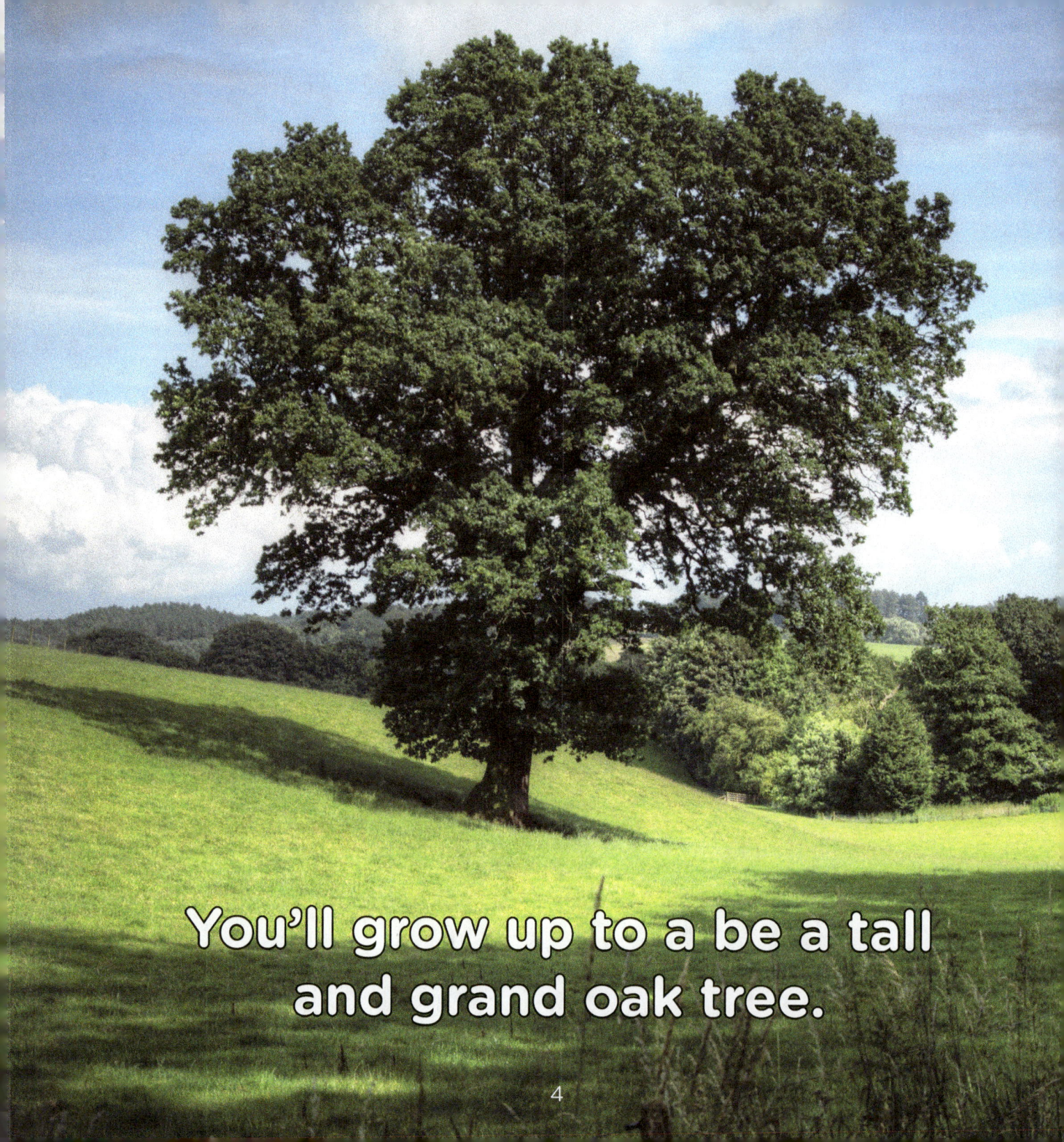

You'll grow up to a be a tall
and grand oak tree.

I wonder what I'll look like when
I grow up.

You'll grow up to be
a graceful paddler:
a quacking duck.

I wonder what I'll look like when
I grow up.

You'll grow up
to be a
roaring lion:
king of the jungle.

I wonder what I'll look like when
I grow up.

You'll grow up to belong
in a forest of ferns.

I wonder what I'll look like when I grow up.

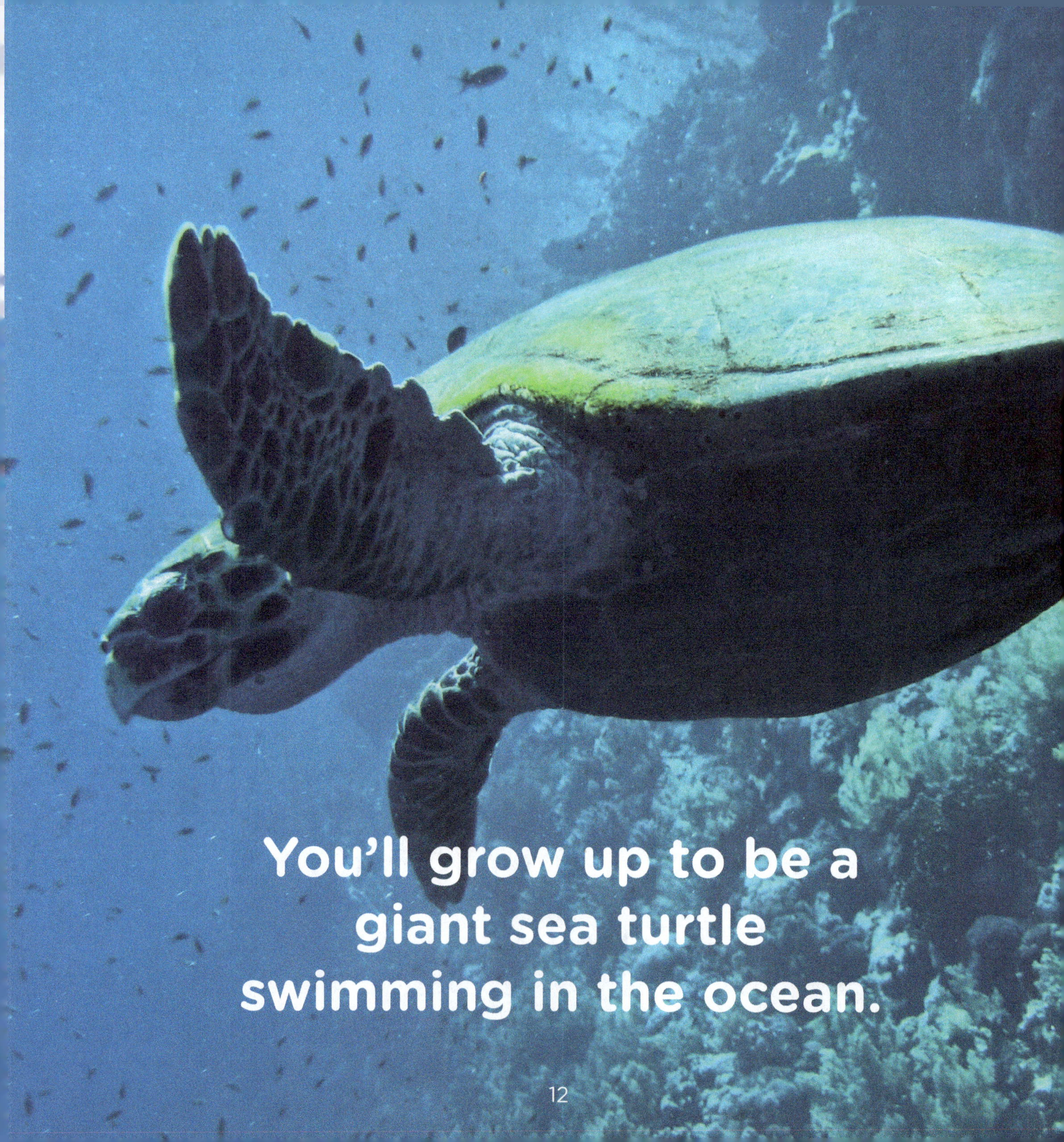

You'll grow up to be a
giant sea turtle
swimming in the ocean.

I wonder what I'll look like when
I grow up.

You'll grow up to be a speedy, white-tailed deer.

I wonder what I'll look like when I grow up.

You'll grow up to be a leaping and croaking frog.

I wonder what I'll look like when I grow up.

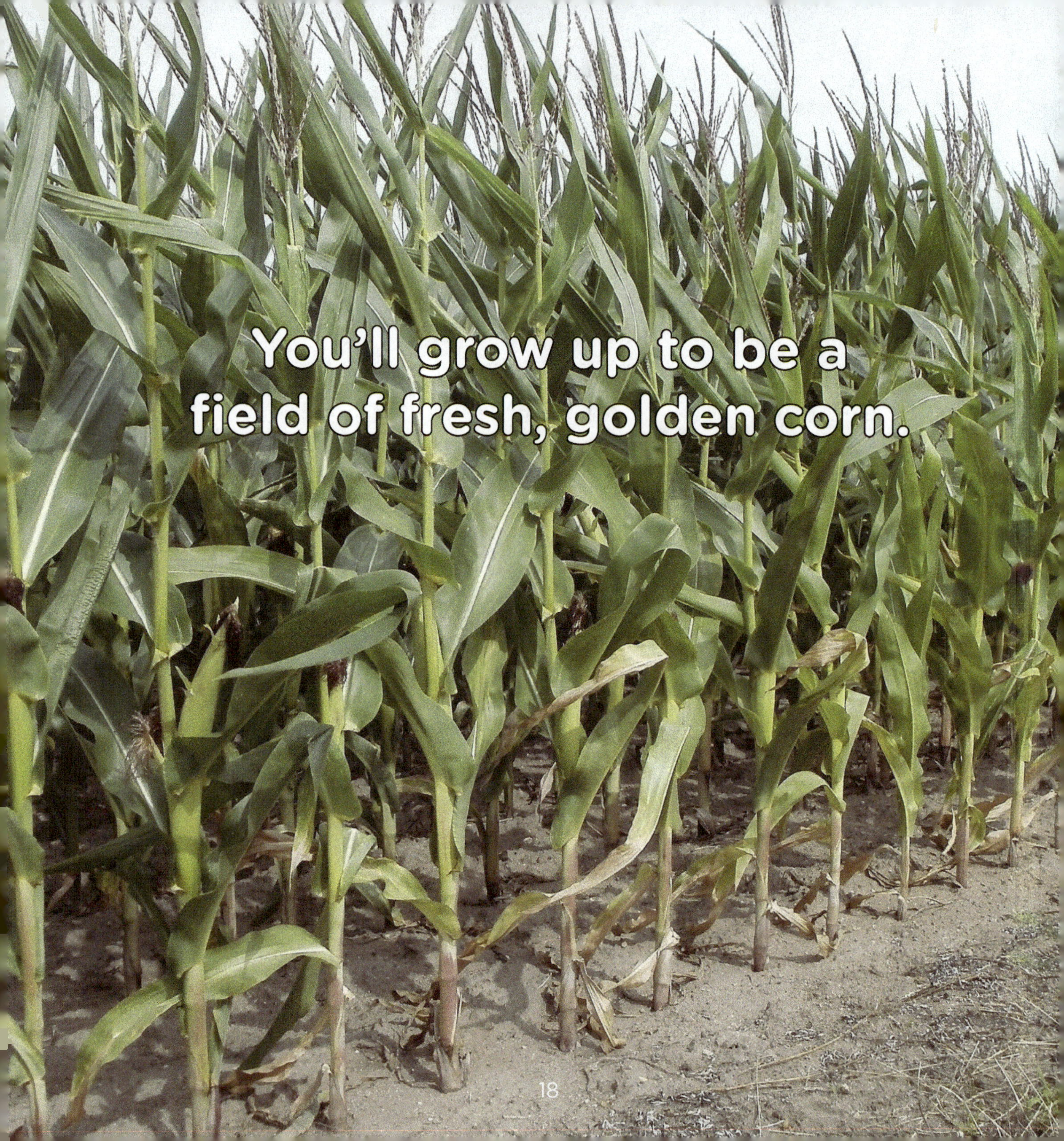

You'll grow up to be a field of fresh, golden corn.

I wonder what I'll look like when
I grow up.

You'll grow up to be a sharp-eyed eagle.

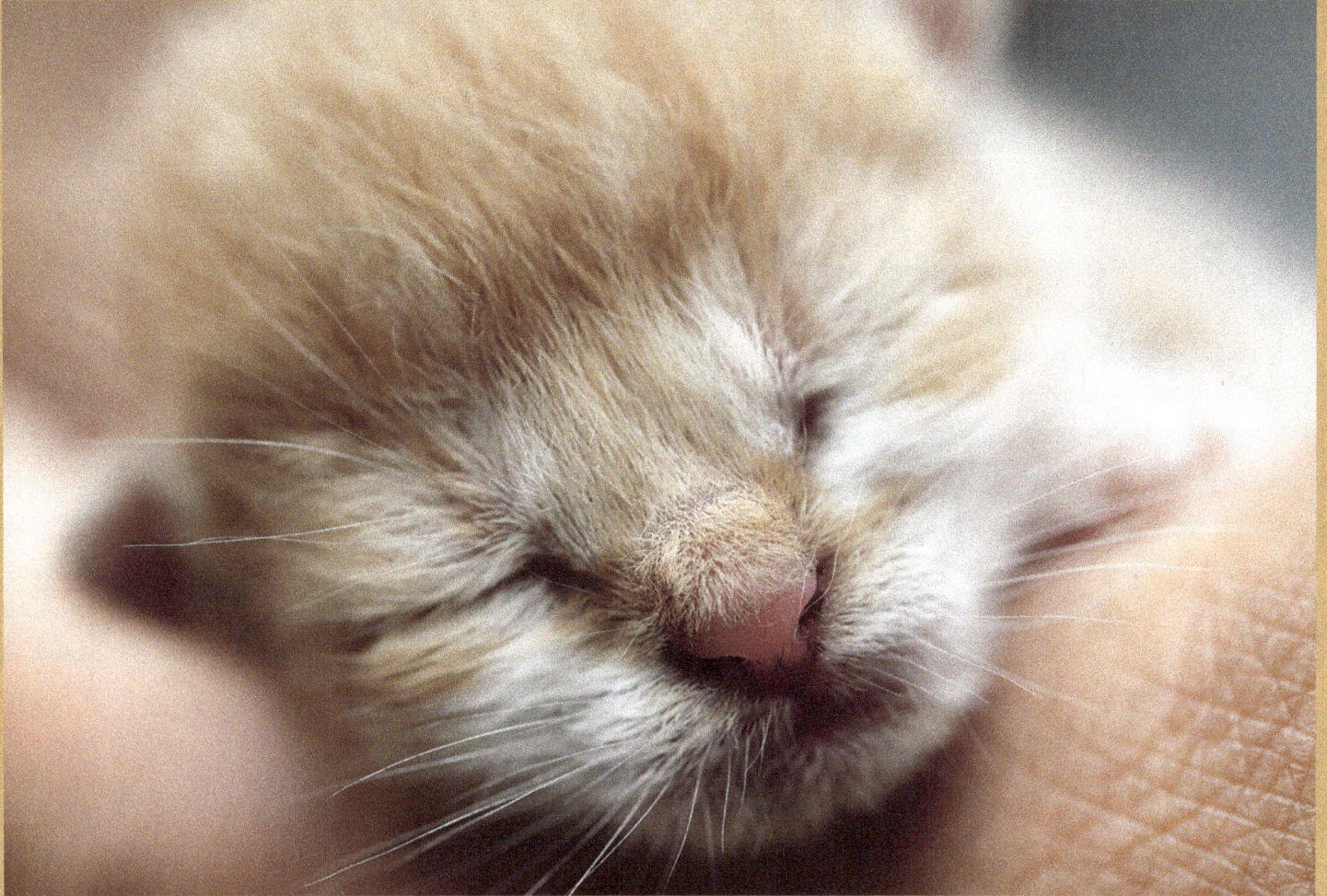

I wonder what I'll look like when
I grow up.

You'll grow up to be a fluffy, prowling cat.

I wonder what I'll look like when I grow up.

You'll grow up to be a prickly hedgehog.

I wonder what I'll look like when I grow up.

You'll grow up to be a sweet and tasty pineapple.

I wonder what I'll look like when
I grow up.

You'll grow up to be a snow white polar bear.

I wonder what I'll look like when I grow up.

You'll grow up to be a penguin that walks on land and swims like a fish.

I wonder what I'll look like when
I grow up.

You'll grow up to be the tallest tree of all!

I wonder what I'll look like when
I grow up.

You'll grow up to be a
fluttering butterfly.

I wonder what I'll look like when
I grow up.

You'll grow up to be a teacher, an engineer, a doctor or just a terrific human being.

# Ozzie & Alina Adventures

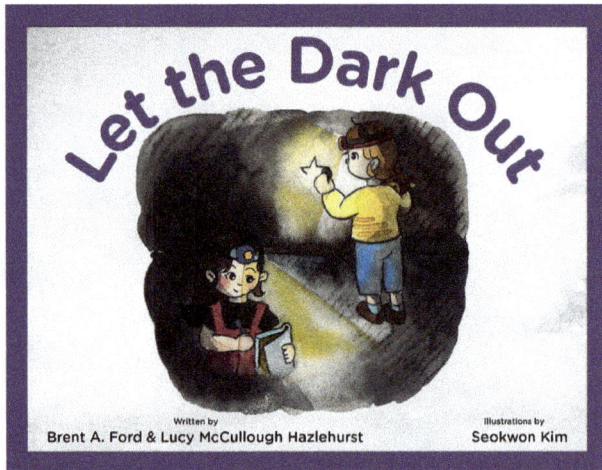

Let the Dark Out

Written by
Brent A. Ford & Lucy McCullough Hazlehurst

Illustrations by
Seokwon Kim

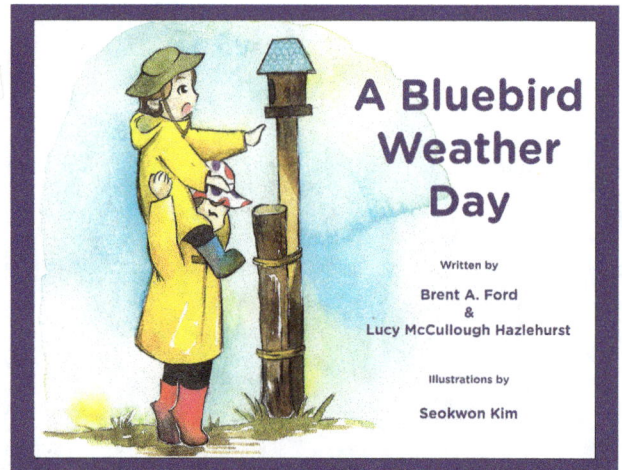

A Bluebird Weather Day

Written by
Brent A. Ford
&
Lucy McCullough Hazlehurst

Illustrations by
Seokwon Kim

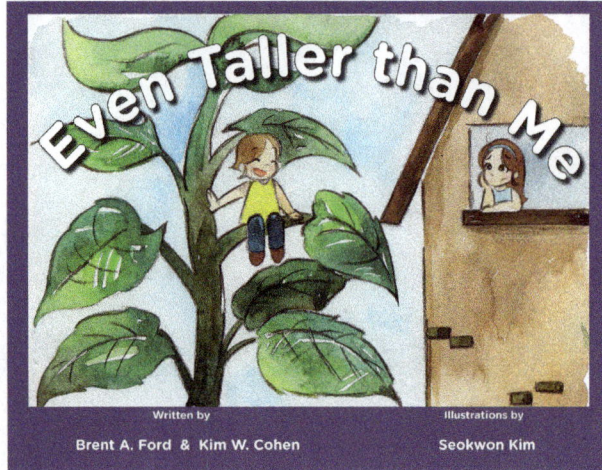

Even Taller than Me

Written by
Brent A. Ford & Kim W. Cohen

Illustrations by
Seokwon Kim

# Updated Classics

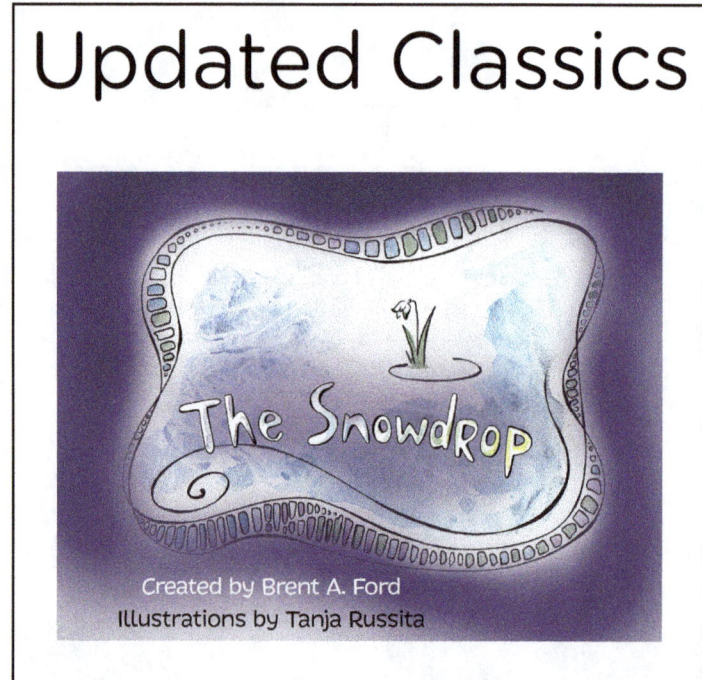

The Snowdrop

Created by Brent A. Ford
Illustrations by Tanja Russita

# Adventures from nVizn Ideas

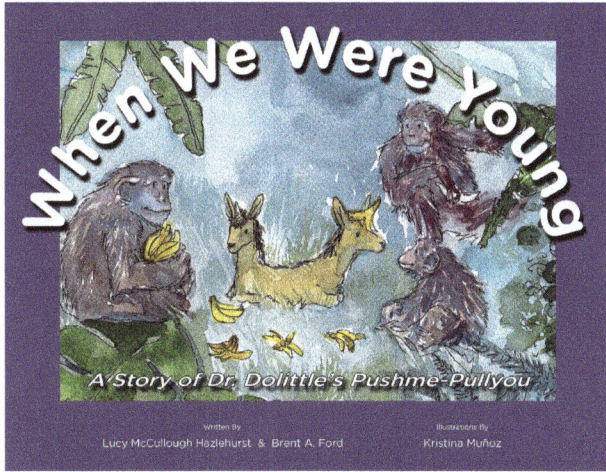
When We Were Young
A Story of Dr. Dolittle's Pushme-Pullyou
Written By Lucy McCullough Hazlehurst & Brent A. Ford
Illustrations By Kristina Muñoz

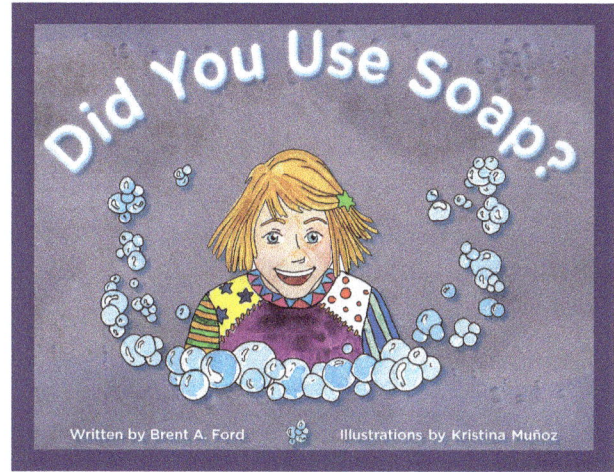
Did You Use Soap?
Written by Brent A. Ford
Illustrations by Kristina Muñoz

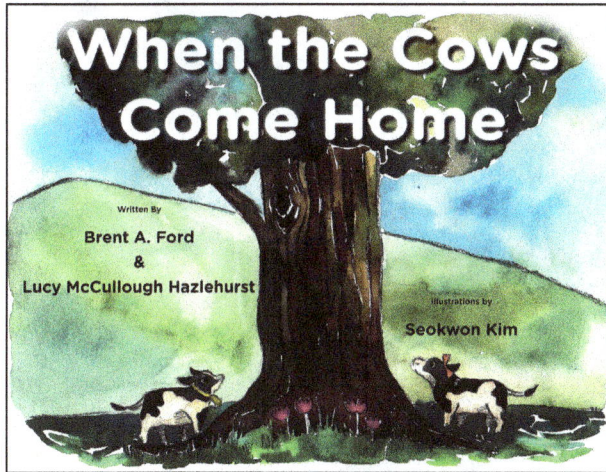
When the Cows Come Home
Written By Brent A. Ford & Lucy McCullough Hazlehurst
Illustrations by Seokwon Kim

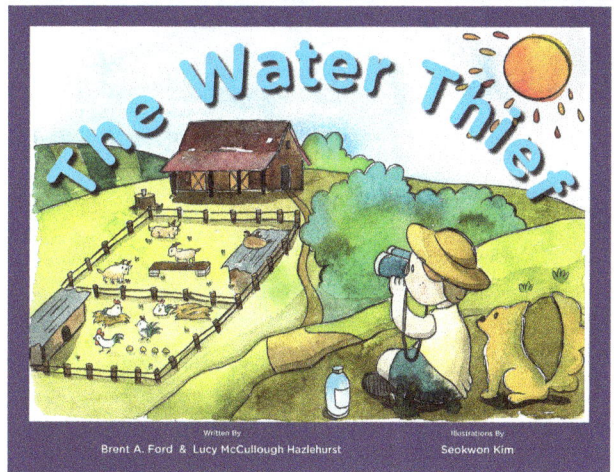
The Water Thief
Written By Brent A. Ford & Lucy McCullough Hazlehurst
Illustrations By Seokwon Kim

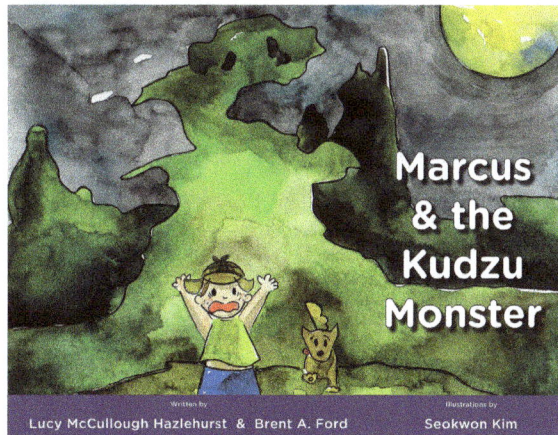
Marcus & the Kudzu Monster
Written by Lucy McCullough Hazlehurst & Brent A. Ford
Illustrations by Seokwon Kim

# Science & Nature eBooks

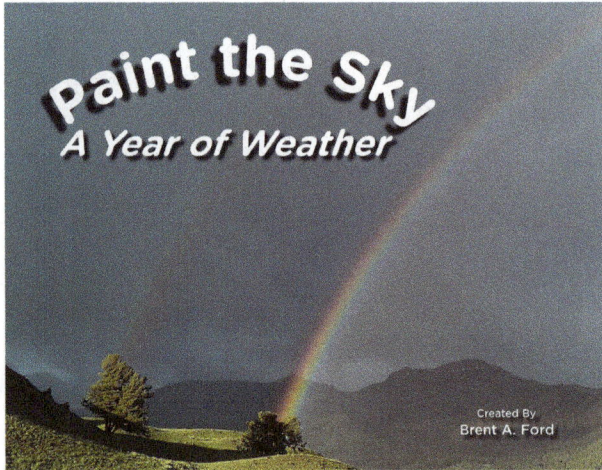

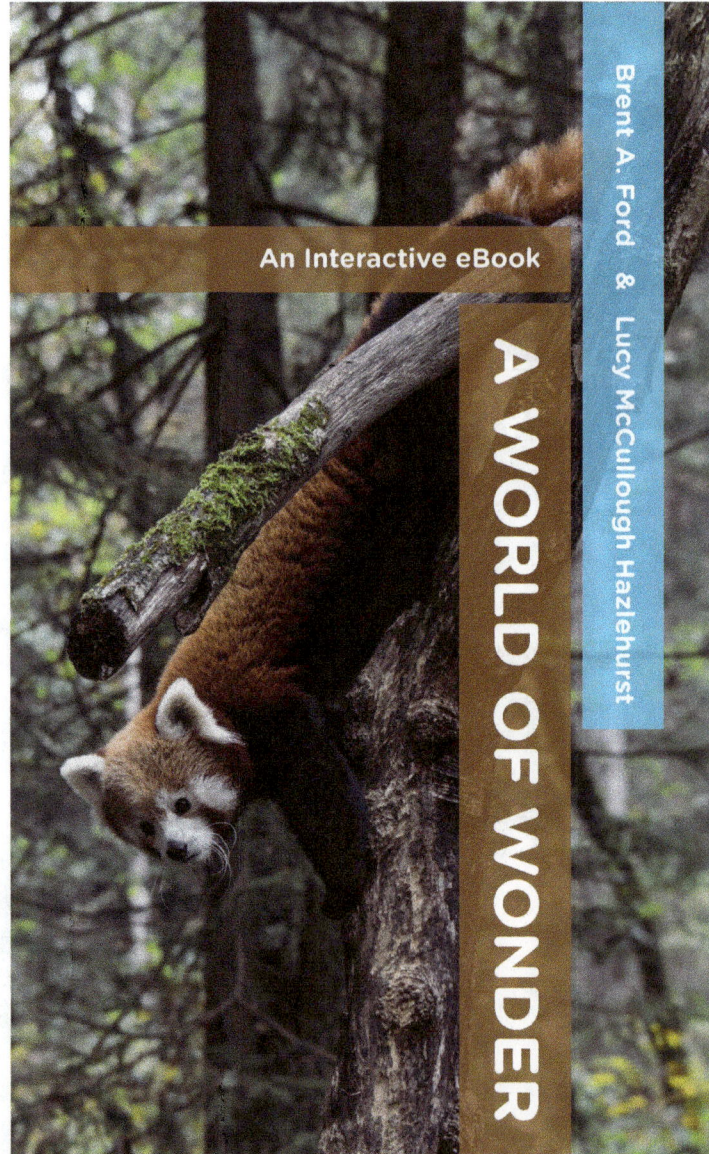

Paint the Sky
A Year of Weather

Created By
Brent A. Ford

An Interactive eBook

A WORLD OF WONDER

Brent A. Ford & Lucy McCullough Hazlehurst

# Interactive eBook

Variety in the Animal World

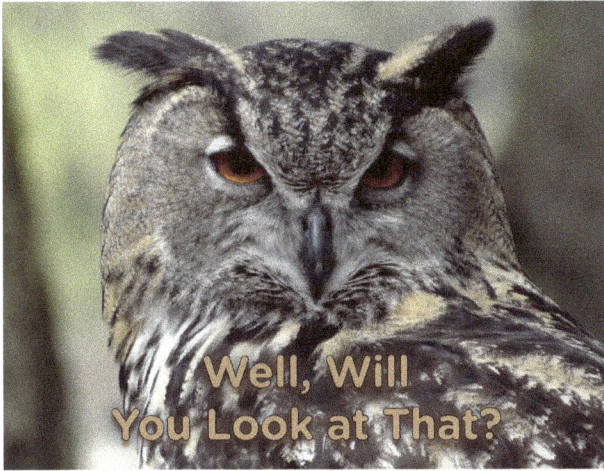
Well, Will You Look at That?

Well, Will You Look at That?

Well, Will You Look at That?

Well, Will You Look at That?

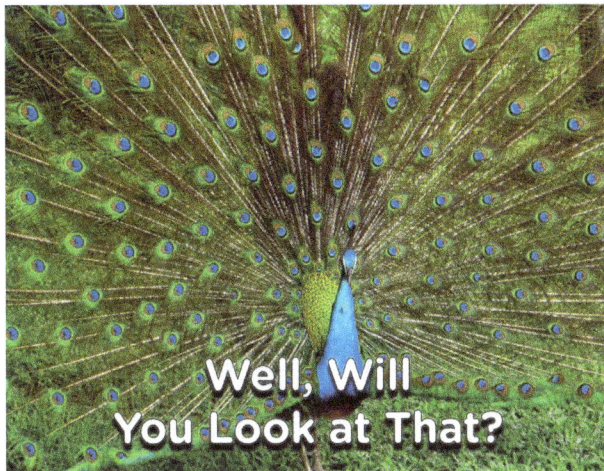
Well, Will You Look at That?

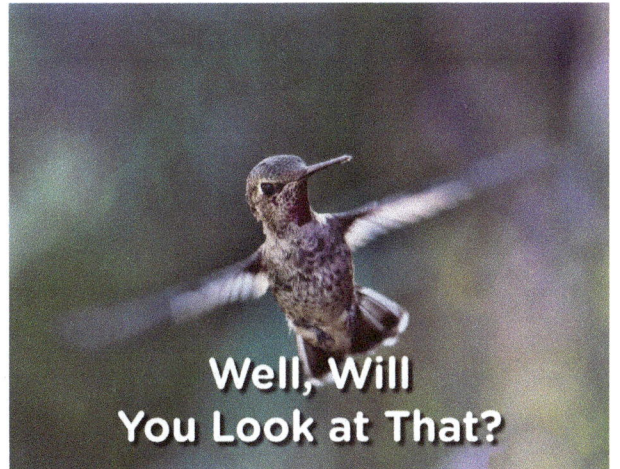
Well, Will You Look at That?

www.ingramcontent.com/pod-product-compliance
Lightning Source LLC
Chambersburg PA
CBHW081148020426
42333CB00021B/2707